YOUR KNOWLEDGE HAS VALUE

- We will publish your bachelor's and master's thesis, essays and papers

- Your own eBook and book - sold worldwide in all relevant shops

- Earn money with each sale

Upload your text at www.GRIN.com and publish for free

Bibliographic information published by the German National Library:

The German National Library lists this publication in the National Bibliography; detailed bibliographic data are available on the Internet at http://dnb.dnb.de .

Imprint:

Copyright © 2017 GRIN Verlag
Print and binding: Books on Demand GmbH, Norderstedt Germany
ISBN: 9783668652224

This book at GRIN:

https://www.grin.com/document/385973

William Fidler

Cosmological Heresies. A new model of the evolution of the universe

GRIN Verlag

Cosmological Heresies - a new model of the evolution of the universe

W M Fidler

Abstract

It is shown in the following work, for a universe obeying Hubble's law and whose vacuum is modelled as a very large set of conjugate pairs of simple frequency-quantised quantum harmonic oscillators, that there is a single mechanism underlying the inception of the universe, the creation of the primary helium, deuterium and lithium, the surface of last scattering and the cosmic jerk. Further, it is demonstrated that these phenomena are not discontinuities but have a finite temporal thickness, albeit, in some of the above, very small.

The oscillators are capable of two modes of operation; the sterile mode, where there is no net energy released into space and which is shown to be the overwhelmingly common state of the universe, and the very rare, but nonetheless highly significant, free vacuon mode, in which elements of energy termed vacuons and of magnitude $\frac{3h\Delta}{2}$, where Δ is the quantisation of the frequency of the electromagnetic spectrum, are released progressively into space.

It is shown that the CMB is not the attenuated radiation emitted from the surface of last scattering, but has its origin in the radiation emitted from the baryonic matter heated at the cosmic jerk, which effect is due to the universe falling into a free vacuon mode

The so-called surface of last scattering is demonstrated not to be the first free radiation observable, in principle in the universe, and which is ascribed in other models to the first clearing of the photonic fog following the Big Bang. It is shown that the first free radiation emanates from a free vacuon mode in which, due to the very high temperatures which may be attained, nuclear reactions occur that could produce the two isotopes of helium, deuterium and lithium (hereinafter referred to as the light elements); this radiation is attenuated significantly before the temporal position of the surface of last scattering is attained. It is noted that, in practice, direct observation of this light element formation event is not possible, for it is 'obscured' by the surface of last scattering. The surface of last scattering is shown to be the result of the universe falling again into a free vacuon mode with the consequent heating of the baryonic matter whose composition at this stage consists of primary hydrogen and that proportion converted previously into the light elements. For simplicity of analysis the baryonic matter here is assumed to be composed solely of diatomic hydrogen.

It is explained that this model has no beginning, that the so-called Big Bang is inconsistent with the model, and that space and time do not have an origin.

It is argued that the concept of dark energy is a consequence of the critical density derived from the Friedmann equations; in this model the creation of space is shown to be due simply to the 'unwinding' of the oscillators, thus giving the impression that the creation of space is caused by an inexplicable phenomenon which has been termed dark energy.

List of contents

Introduction

The surface of last scattering is regarded as a consequence of the Big Bang description of the origin of the observable universe. It is envisaged as a spherical surface from which all of the CMB photons emanate and centred on the Earth. Because of the opacity of the universe prior to the formation of this surface this taken to be the farthest redshift that may be observed.

The discovery of the cosmic jerk was made, independently, by groups led by Schmidt [1] and Perlmutter [2]; the effect is shown clearly in their data as presented by Krauss, p85, [3]. It was found that approximately five billion years bp the expansion of the universe appeared to change suddenly from a state of deceleration to one of acceleration and is attributed to the entity called dark energy. It is not known whether this acceleration will persist into the indefinite future. Indeed, the future of the universe is a subject of much discussion, not to say, controversy.

In this work the vacuum of the universe is modelled as a very large set of conjugate pairs of bi-modal, frequency-quantised, simple quantum harmonic oscillators, one having positive energy, the other, negative energy. Further, each pair of oscillators occupies a cubical enclosure of side equal to the wavelength corresponding to their current frequency, and related through the relationship, $vl = c$.

The magnitudes of the first state energy of each oscillator are equal (and, of course, opposite). We describe, in detail, the modes of behaviour of the system of oscillators and note that the quantum of frequency, Δ, will be discussed at length later.

The first mode is considered to be the 'natural' and overwhelmingly common condition of the universe, as follows:

Each oscillator is considered to be in the first state energy condition, i.e. $^{3hv}/_2$ and $^{-3hv}/_2$, respectively. At some instant the frequency of the positive oscillator falls by the amount, Δ; the oscillator is now at a frequency $v - \Delta$. It is posited that the dimension of the enclosure does not increase at this stage due to the presence in space of a quantity of energy of magnitude, $^{3h\Delta}/_2$, which has the effect of increasing the curvature of space and so inhibits the expansion. The two oscillators are in a state of disequilibrium, for, one is oscillating at frequency $v - \Delta$ whilst the other is oscillating at frequency, v. Equilibrium between the two is restored by the negative energy oscillator emitting a quantity of energy of magnitude, $^{-3h\Delta}/_2$. The two elements of energy annihilate each other, space is now empty, and the pair of oscillators are again in equilibrium and oscillating at frequency, $v - \Delta$. At any time, and in this mode these effects are posited to occur simultaneously throughout the universe, and so all of the elements of the vacuum are now oscillating at frequency, $v - \Delta$.

The above series of processes may be envisaged as being located on a pair of adjacent hyperframes, one larger than the other, for the frequency of the system is now lower than before and hence the size of the set of enclosures, larger, and separated by some finite time interval. Indeed, we could identify the temporal relationship between the hyperframes in this mode by labelling them with the appropriate magnitude of the frequency. We have named the amount of energy, $\pm 3h\Delta/2$, the vacuon, and hence the above sequence could be

envisaged as either the merging of positive and negative vacuons with mutual annihilation, or the exchange of vacuons between the oscillators, in both cases preserving the zero energy condition mentioned earlier. This mode is, for obvious reasons termed 'sterile', for the net energy in space is always zero, but, due to the decrease in the frequency the wavelength increases with the concomitant increase in space.

The second mode is also described at length in [4], and consists of the following sequence of events:

As before, the frequency of the positive oscillator, oscillating at frequency, v, falls by the amount, Δ. The frequency of the oscillator is now $v - \Delta$ but, as before, the expansion of space is inhibited by the presence of the positive vacuon. The positive oscillator now emits a photon of frequency, $v - \Delta$, and the oscillator then falls into the ground state with energy density, ρ_E. It is shown in [4] that in this condition space, treated as a perfect fluid, is expanded by an amount δV through the expenditure of an element of work, $\delta W = -3h\Delta/2$ We now amend the description given in [4]. Since this expansion of space occurs when the positive oscillator is in the ground state then it may be considered that the ground state energy increases by $3h\Delta/2$, for, a decrease in the energy of the ground state by a negative amount is equivalent to an increase in the positive energy of that state. This new energy state is unstable and is restored to its original state by the negative oscillator falling in frequency by an amount, Δ, with the accompanying emission of a negative vacuon of magnitude, $-3h\Delta/2$, which is absorbed by the positive oscillator, returning it to the stable ground state. The initial photon is now reabsorbed and the positive oscillator returns to the first state energy level, $3h(v - \Delta)/2$. The positive and negative oscillators are then in an equilibrium relationship

similar to that which existed before the frequency drop, but now there is an excess amount of energy in space of magnitude, $3h\Delta/2$. Moreover, space has been expanded by the expenditure of mechanical work when the positive oscillator is in the ground state.

The whole series of processes may then repeat.

This mode is called the free vacuon mode; in the history of the universe as modelled here it is taken to be a very rare occurrence and, as shown later, is posited to be the mechanism associated with the inception of the universe, the formation of the light elements, the surface of last scattering and the cosmic jerk.

It should be emphasised that both of these modes do not exist in their pure form with the exception of the period of time, of indefinite extent, before the inception of the universe in the case of the sterile mode, and during the inception phase when the dark and baryonic matters are formed. Thereafter the presence of matter/energy 'contaminates' both modes, but, it is assumed that their actions, as described, are essentially unaffected by this.

Given that the model presented here may be extended indefinitely into the past, then there is no Big Bang at which space and time had their origin---although it may readily be argued that there was a Genesis, in the sense that matter was created from energy, not by the high temperatures of the current orthodoxy, but by the extreme curvature of space caused by the extremely high energy densities pertaining at that time. Moreover, it is shown in [5] that inception occurred when the universe was the size of a cube of side, approximately 4x10^-5 light years; this result is predicated upon there being 10^80 baryons in the current universe [11], and uses the current density proportions of the universe as determined by data from WMAP, [8]. Further, in [5] it is shown that a possible candidate for dark matter consists of new, electrically-neutral structures, the lowest energy of which contains two up quarks and four down quarks, arranged in a ring and having a mass equivalent of close to eleven neutrons. It is posited in [5] that this minimum energy structure is formed first from baryonic matter in the inception phase of the universe, and, from calculations of its approximate dimensions and assuming that the universe was filled with such matter at that time, that the vacuum was oscillating then at a frequency of approximately 10^28 Hz. This sets the upper limit of the frequency of the vacuum which is considered in the following discourse.

The cosmological deceleration parameter, q.

Now, Hubble's parameter, **H** is related to the scale factor, **a**, by: $H = \dot{a}/a$, where the dot indicates differentiation with respect to time.

A dimensionless measure of the deceleration parameter, **q** of the cosmic acceleration of the expansion of space in a FLRW universe is <u>defined</u> by the expression: $q = -a\ddot{a}/\dot{a}^2$.

As shown in [6], Hubble's parameter and **q** are related through: $dH/dt + (1+q)H^2 = 0$.

This may be written: $d(1/H) = (1+q)dt$. In conjunction with equation (4) of [6], viz. $\frac{1}{v}\, dv/dt = -H$, it is shown in [6] that a series of equations relating, respectively to $q = 0$, **q = a constant**, and an assumed exponential distribution of the form, $q = e^{k(t_0-t)} + B$ may be derived, where the subscript **0** indicates the present time. It is the last of these distributions that we proceed to investigate.

All of the series of equations associated with the above distributions have the results, $a_0/a = l_0/l = v/v_0$ in common. For the exponential distribution shown above, the

frequency ratio, which is given by the integration of equation (4), as shown in [6], i.e. $ln\, v/v_0 = \int_t^{t_0} H\, dt$, is complicated by the form of the expression derived there for the Hubble parameter, viz.

$$H = \frac{H_0}{\{1 - H_0[(1 + B)(t_0 - t) + (e^{k(t_0 - t)} - 1)/k]\}} \quad \text{------- (1).}$$

Further, although the magnitude of **0.5**, favoured for **q** by some theorists could be used in the iterative equation (19) of [6] to determine the values of the constants **k** and **B** by setting the age of the universe, $(t_0 - t_1)$ to $2/3H_0$ and the time $(t_0 - t_2)$, at which the cosmic jerk occurs to 5 billion years bp, (equivalent to 0.1577×10^{18} s), see [6], it was considered that this age for the universe, in terms of the currently- accepted age, was unrealistic.

Indeed, it was stated further that the magnitude of **q** is not necessarily equal **to 0.5**. For this reason the iteration equation (21) of [6], **with** $x = H_0(t_0 - t_1)$ was rewritten, as follows:

$$k_{n+1} = \frac{H_0}{x} \ln(q + e^{k_n(t_0 - t_2)}) \quad \text{------------ (2).}$$

As in [6] the procedure for determining the approximate temporal position of $\mathbf{q}$ consists of specifying, as parameter, a value for $\mathbf{q}$ in the range $\mathbf{0 - 0.5}$, then, for a given value for $\mathbf{x}$ the value of $\mathbf{k}$ is obtained by iteration. The magnitude of $\mathbf{B}$ may hence be determined and these values are inserted in the denominator of equation (1). The whole process is then repeated until the magnitude of the denominator changes sign. At this juncture we then, with good accuracy have the values of $\mathbf{k}$ and $\mathbf{B}$ which, in the assumed exponential distribution, will determine the temporal position of $\mathbf{q}$, beyond which equation (1) breaks down, in the sense of yielding non-physical results beyond the singularity where the denominator of (1) becomes zero.

Inspection of equations (1) and (2) show a dependence on the value of $\boldsymbol{H_0}$. Accordingly we now use a magnitude of $\boldsymbol{H_0}$, viz. $\boldsymbol{2.177\ x\ 10^{-18}\ s^{-1}}$, determined from the SDSS – 111 Baryon Oscillation Survey [7], and published on 13 July 2016. It is noted that the Hubble time which corresponds to this value of the Hubble parameter exceeds the currently-accepted age of the universe [3], viz. $\boldsymbol{13.72\ x\ 10^{9}\ yr}$, although, in this model the universe has a time of inception before the present time which may be adjusted by the methods outlined later; the model may be extended into the indefinite past where, before inception, the universe was dark, cold, empty (on a basis of free radiation and matter) but completely determinable in terms of the frequency of oscillation of the vacuum, and the physical size.

As noted earlier, in this model the sterile mode in its pure form only exists before the inception of the universe, for, after inception, space contains matter and radiation even without further emission of free vacuons, and it is proposed that in this condition the variation in the cosmological deceleration parameter is represented by the exponential distribution shown previously.

The time, bp, at which equation (1) breaks down may be set arbitrarily closely to the age of the universe by simply varying the magnitude of $\mathbf{q}$ in equation (2).

Dark energy and this model.

The concept of the critical density of the universe derives from the Friedmann solution of Einstein's field equations for a FLRW metric and a perfect fluid with mass density, ρ, and pressure, p. If the density of the universe is the critical density then, according to the Friedmann solution the geometry of the universe is Euclidean, and hence flat in the sense that there is zero curvature of space.

The space probe WMAP operated from 2001 to 2010 and measured the wavelengths of radiation in the microwave region of the electromagnetic spectrum across the sky. From these measurements it was determined that the universe is flat to a high degree of approximation. It then followed from the Friedmann equations that the actual density state of the universe was very close to the critical density (9.9* 10^-27 kg/m^3). Now, the combined densities of the baryonic matter and the dark matter only account for roughly 29% of the critical. In [1] and [2] it was determined that the universe was accelerating, and it was then inferred that if there exists a positive energy which has the effect of exerting a negative pressure and hence a gravitationally-repulsive effect on space, then, if this energy has a density, 71% of the critical, then this would account for the difference. Hence, it was posited that there actually existed such a form of energy in space. Whilst this rendered the overall density in space equal to the critical, thus satisfying the observed condition of flatness, it immediately raised the question of the nature of this dark energy.

To date there has been no definitive explanation of dark energy. Further, it is remarked, with astonishment, by various authors that the fact that the current magnitude of the density of the universe is virtually that of the critical, is akin to a pencil balancing on its point. It is contended that this should at least instigate further investigation of the equations upon which the assertion of critical density is founded.

We now offer the heretical suggestion that the notion of critical density is a consequence of the Friedmann equations and that the universe, as modelled here, is flat because the rate at which space is being created gives the impression that a mechanism exists which has a positive magnitude whilst, at the same time is gravitationally-repulsive, but is, in fact, due to the 'unwinding' of the oscillators described earlier.

We offer no explanation for what wound up the oscillators, but since our model may be extended into the indefinite past it is considered that the requirement for an explanation does not arise.

The hypothesis of minimum curvature.

Flatness of space is guaranteed by the complete absence of matter and/or radiation. Prior to the inception of the universe, space in the model presented here was absolutely flat, since there it was in the condition referred to earlier as the sterile, or natural mode where the net energy, attributable to all sources in space, was zero. It is then averred that space obeys a law, which we propose to call the hypothesis of minimum curvature, and which may be stated as follows: space will respond to the presence of matter and/or radiation by expanding such that the densities of these deforming entities are minimised. In the limit, space will become flat when these densities are reduced to zero. The principle could be articulated more succinctly by the statement 'space abhors mass'. If we use the properties of shape-memory materials (e.g. Nitinol) as analogy, then the idea of shape memory could be applied to space. It may then be posited that the remembered equilibrium curvature of space is zero and that the universe, as constituted, represents a deviation from this state of equilibrium.

Frequency quantisation of the electromagnetic spectrum

It was shown in [9] that the whole of the electromagnetic spectrum, as far as the Planck frequency could be rendered dimensionless and contained within a right-angled isosceles triangle of dimensions, 1,1, $\sqrt{2}$; moreover, the triangle was shown to be the projection of a three dimensional helical surface. In addition, it was determined that there existed elements of the geometry of the triangle which formed a fractal path and from which it was shown that the frequency could be quantised, according to the formula:

$$\Delta = {(V)}\Big/{(\sqrt{2})^{n-1}}$$

In this formula, Δ is the magnitude of the frequency quantisation, (V) is the Planck frequency, and $\mathbf{n}$ is a disposable odd integer.

Because of the nature of a fractal the frequency quantisation could be made arbitrarily small but never zero. In addition, it was shown, [10] that there was no prohibition regarding the incorporation of the concept of frequency quantisation into a simple quantum harmonic oscillator.

It is posited that a drop in frequency of amount, Δ , from any given frequency, v, will engender an increase in the volume of space occupied by an oscillator, which increase is independent of the process with which the frequency drop is associated. Thus, for the same frequency drop the increase in the volume of space is the same for the system of oscillators in both the sterile and free vacuon modes. Whether frequency quantisation is real or simply a convenient concept which permits counting, it allows the avoidance of the nebulous, differential quantity.

Some preliminary calculations

We contend that during the period of the WMAP measurements, and, as will be argued later, for a considerable period of time before that, the vacuum was almost wholly in the sterile mode. However, since the Friedmann model requires a dark energy having the characteristics of a positive energy which is gravitationally-repulsive and of magnitude approximately 71.4% of the critical density, and we also employ the dark and baryonic matter densities in this and previous analyses, we take advantage of the concept of dark energy and ascribe its density to be that of the ground state energy density, ρ_E of the positive energy oscillator, for this has a positive energy, but, as noted by Davis and Griffen [17] is gravitationally-repulsive; the equation of state for space modelled as a perfect fluid has the form, $P/\rho_E = -1$. Thus, the ground state energy density is taken to be 71.4% of the critical and this is used to calculate the current frequency of the oscillators.

Although most of the following calculations have been done elsewhere, for completeness, and using the explanation given above, we now repeat them.

The ground state energy density, ρ_E, of the positive oscillator is given by:
$\rho_E = 0.714 * \rho_C$, where, ρ_C, the critical density, has a magnitude of $9.9 * 10^{-27} \ kg/m^3$.

For a cubical enclosure, $\rho_E = h\nu_0 \big/ 2c^2 l_0^3 = h\nu_0^4 \big/ 2c^5 \ \rightarrow \ \nu_0 = 2.681 \ THz$, where the

subscript '0' denotes the present time.

The side dimension, l_0 of an enclosure at the current time, is given by:

$$l_0 = c/\nu_0 \ \rightarrow \ 1.118 * 10^{-4} \ m.$$

Now, at the current time the mass densities of the baryonic and dark matters are $0.4554*10^{-27}$ and $2.376*10^{-27}$, kg/m^3 , respectively. Roughly speaking this means that

there is one proton in every four cubic meters of space and the 'proton' equivalent of three dark matter structures in every two cubic meters of space. It is then contended that, at the current time these entities occupy space as if the others were not present.

Hence, we may determine the volume, V, of a universe with N_b baryons as follows:

The current ('0') baryonic matter density of the universe is : $N_b m_p / V_0$, where, m_p is the mass of a proton $(1.6725*10^{-27} \text{ kg})$; now, $N_b m_p / V_0 = 0.4554 * 10^{-27}$. Taking the number of baryons in the universe to be 10^{80} as assumed by Penrose, [11], the current volume, V_0, is then calculated to be $3.673*10^{80} m^3$, equivalent to a cube of side $75.75*10^9$ **light years**.

In [6], p16 it is shown that the Hubble parameter, **H**, may be interpreted as one third of the volumetric strain rate (of either an enclosure, or, the universe), i.e. $H = \frac{1}{3V} \frac{dV}{dt}$. Hence we calculate that, currently, space is being created at a rate of, $2.402 * 10^{63} \ m^3/s$;

this should be contrasted with the current volume. The ratio of the latter number to the former yields the current volumetric strain rate, $6.54 * 10^{-18} \ s^{-1}$; from the above it is seen at the current time, that the rate at which space is being created is utterly minute when compared with the volume of the universe.

We may also calculate the number, N_{enc} of enclosures; this number remains constant, the enclosures only varying in size throughout the entire history of the universe and into the indefinite past before inception.

Now, $V_0 = N_{enc} \ l_0^3 \rightarrow N_{enc} = 2.628 * 10^{92}$.

Cosmological Heresies

Heresy 1

As mentioned earlier, we posit that there was no Big Bang, for it is argued later that there was a Genesis in the sense described before, when the space that was to become the universe that is currently observed attained the size of a cube of side $4*10^{-5}$ light years, approximately. Further, the inception of the universe is not the beginning of space and time, for, according to the model presented here the universe has existed, essentially forever. The inception of the current universe is attributed to the vacuum falling into the free vacuon mode for the first time in its entire history.

Heresy 2

The surface of last scattering is the result of the vacuum, for the third time (see later), falling very briefly into the free vacuon mode shortly after the inception of the universe; the energy density consequently arising in this mode being sufficient only to heat the baryons produced during the second matter-production part of the inception phase to the temperature, ascribed in [12] to that at the surface of last scattering, i.e. 3000K. Whilst it is acknowledged that the baryonic matter here is a mixture of hydrogen and the light elements mentioned previously, it is assumed, entirely for the sake of simplicity of the analysis that this matter is composed solely of diatomic hydrogen. It is argued later that, at a time significantly before the surface of last scattering there was an earlier free vacuon mode where the temperature attained by the baryonic matter was high enough for the formation of the light elements to occur by the process of nuclear fusion; in [5] this was identified as the inception of the universe, but is now considered to occur after the creation of the primordial hydrogen from baryonic matter. It is appropriate to call this phase the Big Light-up. The temporal positioning of the surface of last scattering may be moved closer to the time of the inception of the universe by further refining of the calculations of its position. We may, with a measure of confidence state that at the surface of last scattering the matter that was heated consisted only of hydrogen and the light elements, for its temporal position may be located significantly before that of the formation of the first stars.

Further, since the formation of the light elements from some of the primordial hydrogen by the process of nuclear fusion occurs at a time somewhat later than the inception of the universe, it may be implied that the primordial hydrogen was formed, not during some event, when the temperature was enormously large but, as argued later, by the extremely high energy densities generated during the first free vacuon phase, resulting in the curving of space to the extent that it becomes an enclosure, entrapping energy within.

Heresy 3

Approximately five billion years before the present time (the actual time is immaterial) the vacuum, for the fourth time, briefly (cosmologically-speaking) entered the free vacuon mode. As before, the energy density generated was sufficient to heat the baryonic matter there, resulting in a rise in temperature in excess of that of the attenuated radiation from the surface of last scattering and which, at the time of the jerk, is shown to be very small. The CMB measured at the present time is posited to derive almost wholly from the attenuated radiation from the matter heated during the cosmic jerk.

Thus, we attribute a common mechanism to the inception of the universe, the formation of the light elements, the surface of last scattering, and the CMB associated with the cosmic jerk.

It is emphasised that the dark matter, being thermally-inert takes no part in any of the heating processes. Further, free vacuons are never encountered at the present time for they are the source of energy in the production of dark and baryonic matter at the inception of the universe, and in the heating of baryonic matter thereafter; no such heating is posited to have occurred or been detectable for at least five billion years before the present time.

As a working assumption it is proposed that the initiation of all of the free vacuon modes noted above is explained by the catch-all, 'quantum fluctuations'.

A diagrammatic representation of the variation of the cosmological deceleration parameter, q, with time, from the inception of the universe' 0', to the present time.

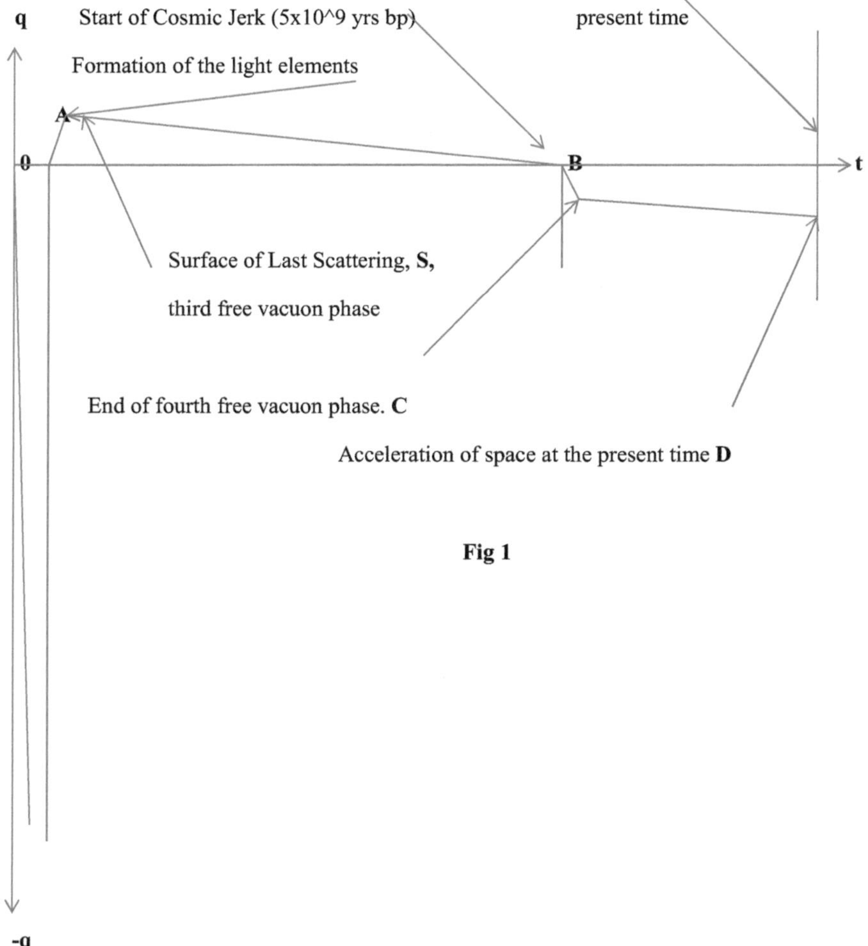

Fig 1

Fig1 is a simplified diagram showing the temporal variation, as proposed in this work, of the cosmological deceleration parameter, q, from the inception of the universe up to the present time.

The diagram, which is hugely out-of-scale, is solely for the purpose of illustration and is presented as a series of straight lines.

It was shown in [6] that the linear acceleration (or deceleration) of space, dv/dt, is given by :

$$dv/dt = -c/2 \sqrt[3]{N_{enc.}} \left(qH^2/v \right) \text{------------ (3).}$$

This is entirely independent of any expression which describes the trajectory of the deceleration parameter. It is readily seen that the states of acceleration and deceleration are represented, respectively, by negative and positive magnitudes of **q**.

Point **O** represents the inception of the universe at a temperature of absolute zero, followed immediately by a phase of very large acceleration (**-q**). This is taken to be a free vacuon phase in which quarks, gluons and electrons are formed from the vacuons, not by some condensation process, but by the extreme curving of space. Further, it is assumed that all of the energy released in the formation of the fundamental entities is stored within their structure; there is no excess radiation, and hence the temperature remains at absolute zero. The release of free vacuons continues and this energy is completely absorbed in the process of the assembly of baryonic matter from the quarks and gluons. The investigation of how the vacuons may be associated with the structure of the quarks, gluons and electrons, using the approach developed in [6] to describe how space may be endowed with structure from an assembly of identical entities, is an endeavour postponed to another day.

The positive baryons capture the electrons produced earlier and form the simplest and most numerous structure, hydrogen. Space is now almost completely filled with baryonic matter. The density is such that the electrons are then forced down onto the nucleus resulting in any protons being converted into neutrons; it is assumed that the mixture is now composed solely of neutrons; further transition is resisted by neutron degeneracy pressure. Now, Montgomery and Jeffrey [13] modelled baryons as triangular ovoids, whilst Llanes- Estrada and Navarro [14] showed that, under conditions of extreme pressure the shape of neutrons may be changed progressively from spheres into cubes. These shapes violate the spherical shape envisaged for baryons in the Standard Model of Physics. We have modified the ideas in [14] to suggest that the triangular ovoids of [13] may be deformed into two-dimensional equilateral triangles under the extreme conditions existing in the first part of the free vacuon phase at the start of the universe, and it is these deformed ovoids which may be assembled into dark matter by the processes described in [5]. Moreover, the energy of formation is, as before, stored in the structures; the lowest energy structure of this type, formed from two deformed neutrons, is shown, [5], to have a mass equivalent of approximately eleven neutrons. The temperature remains at absolute zero, for all of the energy is stored within the structures and there is no excess radiation. We abandon the notion that high temperatures are necessary at the inception of the universe for the formation of matter from energy, for we have shown previously, [5], that, for this model, the ratio of the rate at which energy is entering space relative to the rate at which space is being created is equal to the ground state energy density of the positive energy oscillator and, during inception the calculation below shows that this ratio is very large. It is assumed that this 'cramming' of energy results in such an intense curvature of space that some of space locally closes, entrapping energy within it. We must not impute the thermodynamic notion that what is essentially compression of energy has the same thermal outcome as compression of matter. Indeed, the dynamics of space and energy are embodied, not in Thermodynamics, but in the theories of Special and General Relativity. Further, in the light of the curvature of space hypothesis advanced earlier it may be envisaged that the gravitational attraction of the matter 'trapped' within a closed region of space overwhelms the tendency of space to seek the condition of flatness, thus maintaining the integrity of what may now be considered to be, a particle, whose mass may be derived from Einstein's
$E = mc^2$.

The frequency of the oscillators at inception, v_i was determined, [5], to be approximately $0.5*10^{28}$ **Hz**. The mean ground state energy density, $\rho_{eg,}$ during the matter formation interval may then be determined roughly from the formula, $\rho_{eg} = {}^h/_{4c^3}\, v_i^4$, which, for the above frequency has the magnitude: $0.58 * 10^{51}\, {}^J/_{m^3}$.

This may be interpreted as $0.58 * 10^{51}$ **(J/s)/(m^3/s)** , thus substantiating the statement made above. For comparison, using the magnitude of the vacuum frequency at the cosmic jerk determined later (**3.953 T Hz**), the corresponding mean ground state energy density there is $15*(10^{-10})$ **(J/s)/(m^3/s)**, indicating that space here is being created at a far greater rate than the rate at which energy is entering space and, in the light of the explanation above, that the absence of such conditions render the formation of particles, impossible.

Further, during a sterile mode the ground state is never encountered. It then follows that the rate ratio quoted previously is always zero, for whilst space is being created at a particular rate, the rate at which energy is entering space is zero.

The dark matter formation period is followed immediately by another free vacuon phase, of frequency interval **0.064*10^12 Hz**, [5]. The amount of energy released during this interval is sufficient only to produce the very lightest atoms. It is a matter of speculation that the presence of dark matter, to some extent inhibits the release of vacuons, and so limits the extent of the second free vacuon phase. The temperature again remains at absolute zero. During this phase the dark matter interacts with the baryonic matter only through the agency of gravity, and we conclude that, as the dark matter is thermally inert, then there can be no absorption of the energy released during this phase for there is insufficient available to do other than form the currently-measured proportion of baryonic matter. As noted in [5] the dark matter has no electrons and hence no chemistry and no photons. Thus, the detection of this substance poses considerable challenges for the experimenter. Since the dark matter can only respond to a gravitational field, then this is the only mechanism which can accelerate it to sufficient energies to produce the collisions between other dark and/or baryonic matters which might result in its disruption with the accompanying ejecta. It is also suggested that there is the possibility of disruption by collisions with baryonic matter travelling at relativistic speeds; in the context of this work there is no requirement to explore further these speculations.

Through the agency of gravity the dark and baryonic matter progressively reduce the acceleration of space to zero; for the second time the trajectory of **q** intercepts the **q = 0** axis, becoming positive, indicating that the space is now expanding at a decelerating rate.

The deceleration increases in an unspecified fashion until the maximum deceleration is attained at point **A**. By the method described earlier we may position point **A** arbitrarily close to the inception point at **O**. Later, it is argued, that by the process of nuclear fusion, initiated by the vacuum falling briefly into the free vacuon mode from **A**, some of the hydrogen is transformed into the light elements. This phase forms a 'bridge' between the trajectory terminating in the maximum positive value of the deceleration parameter and the exponential trajectory, which, later, we name the Critical Trajectory.
The surface of last scattering is located at **S,** which is in the future of the 'bridge' emanating from **A**; the trajectory of the deceleration parameter between **A** and **B** is formed by an exponential distribution of the type described earlier, and named above.

The cosmic jerk is initiated at point **B**. The change in the trajectory of the deceleration parameter is again due to the vacuum falling into the free vacuon mode. The energy density attained during this process being sufficient only to heat the baryonic matter present to a temperature in excess of that of the attenuated radiation which emanated from the surface of last scattering and the light element formation phase.

It is assumed that the trajectory of the deceleration parameter from the end of the jerk phase until the present time is described by an exponential distribution of the same type as that used previously. In this model the CMB is attributed to the attenuated radiation emanating from the baryonic matter heated during the jerk phase, plus the attenuated radiation emitted from the sources described previously.

Energy density and blackbody radiation temperature

As noted in [5], the energy density in radiation, ρ_r, is given by the formula:

$$\rho_r = a_{rad}\,\theta^4 \text{ ---------------- (4).}$$

Here, θ is taken to be the temperature of a blackbody emitting the radiation and a_{rad} is the radiation constant, of magnitude, **7.565767*10^-16 J/((m^3).K^4).**

The radiation referred to in equation (4) is composed of photons.

We now' borrow' this expression and relate it to the situation where a collection of vacuons accumulates within an enclosure, for such will constitute an energy density.

The energy density under this circumstance is given by $\rho_e = \dfrac{3h}{2}\,(v_i - v_f)/l^3$. The subscripts 'i' and 'f' refer to the initial and final release of a free vacuon. This notation will be used in all future, relevant equations.

Further, we may write $\rho_e = \dfrac{3h}{2c^3}\,(v_i - v_f)v_f^3$.

If we now regard these two energy densities to be equivalent then we may write:

$$\theta^4 = \frac{3h}{2a_{rad}c^3}\,(v_i - v_f)v_f^3.$$

This may be written: $\quad \theta^4 = \bar{F}(v_i - v_f)v_f^3 \text{ ------------- (5).}$

Where $\bar{F} = \dfrac{3h}{2ac^3}$, and has magnitude **4.87653*10^-44 (Ks)^4.**

In [5] this was denoted as **F**, but, in the SI system this symbol is reserved for Faraday's constant.

The surface of last scattering

The mass, M_u of all of the baryonic matter in the universe may, for a universe containing N_u baryons be written: $M_u = N_u.m_p$, where m_p is the mass of a proton **(1.6726*10^-27 kg)**. For the purpose of simplicity, in the following, we assume the mixture to be composed solely of diatomic hydrogen.

The energy, Q_u that must be supplied to raise the temperature of this mass of hydrogen by an amount $\delta\theta$ is given by: $Q_u = M_u.C.\delta\theta$.

If it assumed that the heating process takes place essentially at constant volume, then **C** is the specific heat capacity at constant volume, C_v. Now, if the temperature of diatomic hydrogen is raised to **3000K**, which is approaching the dissociation temperature of approximately **3200K**, then the vibrational modes of the molecule will be almost fully excited. The specific heat capacity at constant volume of any diatomic molecule experiencing vibrational excitation can be shown to be given by: $C_v = {}^{7R_{un}}\!/_2$, where R_{un} is the Universal Gas Constant, of magnitude, **8.3143 KJ/kg-mol K**.

Hence, for diatomic hydrogen at **3000K**, C_v is calculated to be **14.55 KJ/kg K**.

Since we show later that the temperature of the matter just before the surface of last scattering is close to absolute zero then we may replace $\delta\theta$ by θ, the temperature relative to absolute zero.

In a free vacuon mode the energy released into each enclosure where the frequency falls by an amount, δv is given by ${}^{3h.\,\delta v}\!/_2$. The total energy, Q_t is then, $N_{enc}.{}^{3h\delta v}\!/_2$. Equating Q_t to Q_u gives the frequency drop during the heating interval, thus

$$\delta v = {}^{2N_u\,m_p\,C_v\,\theta}\!\big/_{3hN_{enc}} = 28\,Hz.$$

Here, it has been assumed that there are **10^80** baryons in the universe.

From equation (5) we may now calculate the frequency of the vacuum at which the heating process associated with the surface of last scattering, terminates, as follows:

$$v_f = \sqrt[3]{[{}^{\theta^4}\!/_{\delta v\,\bar{F}}]} = 3.9\text{*}10\text{^}18\ Hz.$$

In view of the size of the frequency drop calculated earlier, then the frequency of the vacuum at which the process initiating the heating of the hydrogen begins may also be taken to be **3.9*10^18 Hz,** for the 'width' of the region of last scattering is **28 Hz.**

The temporal trajectory of the deceleration parameter between a point close to the inception of the universe and the cosmic jerk.

The constants in the exponential distribution assumed for **q** may be determined by solving equation (1) in conjunction with equation (2).

By progressively reducing **q** from a value of **0.5** it is found that **q = 0.286** may be 'positioned' **13.715*10^9** years before the present time; this may be contrasted with the currently-accepted age of the universe of **13.72*10^9** years to two significant figures, as noted by Krauss [3]. By extending the magnitude of the deceleration parameter to more significant figures than that shown above, the analysis may be further refined by numerical experimentation thus bringing the above position and the age of the universe into closer correspondence.

The constant, **k**, determined by the procedure described above has magnitude **0.8160679*(10^-18)** s^{-1}, whilst the constant, **B**, which, of course is given by $B = -\exp(k * 0.1577)$, has magnitude -1.13734194, and is dimensionless. The time to the cosmic jerk, Δt_j assumed to occur **5*10^9 yr** from 'now 'may be written as **0.1577*10^18 s** and hence the product $k. \Delta t_j$, may be written explicitly as the product of **k** and. Δt_j, omitting the exponents.

Since the frequency interval of the surface of last scattering has been shown to be very small, then it is assumed that the trajectories of the deceleration parameter before and after the surface of last scattering are virtually the same.

We call the trajectory of the deceleration parameter defined by the exponential distribution with the above-noted constants, the Critical Trajectory, for it may be shown, at **13.715*10^9** years before the present time along this trajectory, that according to equation (1) the frequency of the vacuum becomes infinite.

This result must be regarded with great caution, for in this model the inception frequency of the vacuum is finite and the above statement may be considered as an aspect of an overwrought theory and will be discussed at length later.

The frequency of the vacuum at the beginning of the cosmic jerk

Now, we have determined the frequency of the vacuum, v_s at the surface of last scattering and, if we could determine the vacuum frequency, v_i at the beginning of the cosmic jerk then solving equation (1) for a range of times before the cosmic jerk will, in principle, yield the time interval to the surface of last scattering, for the LHS of the integral expression

$I = {}^{v_s}/_{v_0} = exp[\int_{t_s}^{t_0} H\ dt]$ is then known.

The integration is performed between 'now' and the time of the surface of last scattering, and hence the value of the integral between 'now' and the jerk must be subtracted from the result. It should be emphasised that the v_0 referred to above is not that calculated at the current time, and, indeed, is unknown; although, it is assumed that the current magnitude, H_0 of Hubble's parameter remains constant in all of the analyses.

The constants k and B remain as before, for the trajectory before the jerk is not dependent upon that after the jerk. However we must not use the trajectory defined by the same constants after the jerk to determine the vacuum frequency, v_f at the termination of the jerk, for, as Fig 1 shows clearly, the trajectory between the end of the jerk and 'now' is not that which would obtain from the simple extension of the trajectory from the surface of last scattering to the present time.

Now, ${}^{v_s}/_{v_0} = {}^{v_s}/_{v_i}\ {}^{v_i}/_{v_0}$ and hence, $ln\ {}^{v_s}/_{v_0} = ln\ {}^{v_s}/_{v_i} + ln\ {}^{v_i}/_{v_0}$

Therefore, $ln\ {}^{v_s}/_{v_i} = \int_{t_s}^{t_0} H\ dt - \int_{t_i}^{t_0} H\ dt.$

It is noted in [5] that the integration of the functions above presents a considerable challenge. In all probability the functions are not analytic and quadrature is then only possible by numerical means. This, in itself poses a problem, for the denominator in the expression for H changes sign at the limit of applicability of equation (1), thus passing through zero, the integrand at this point becoming infinite.

Numerical integration of equation (1) may be facilitated by the process of substitution, as follows:

$$\int_t^{t_0} H\ dt = \int_t^{t_0} H_0\ dt \Big/ \{1 - H_0[(1 + B).(t_0 - t) + (exp(k * (t_0 - t) - 1)/k)]\} \quad \text{----- (6)}.$$

Let $\eta = (t_0 - t)/\Delta t_x$, where Δt_x is the interval of time between 'now' and some particular event in the past. Hence, $d\eta = -dt/\Delta t_x$

Equation (6) may now be written:

$$\int_t^{t_0} H\, dt = \left. \int_0^1 H_0\, \Delta t_x\, d\eta \middle/ \{1 - H_0[(1 + B).\eta\Delta t_x + (exp(k\eta\Delta t_x) - 1)/k)]\} \right. \quad \text{-------- (7)}$$

Let, $y = k\eta\Delta t_x$, hence $d\eta = dy/k\Delta t_x$. Under this transformation and with some rearrangement, the above equation becomes:

$$\int_t^{t_0} H\, dt = \left. \int_0^{\hat{y}} (H_0 \cdot {}^1\!/_k) \middle/ \{1 - (H_0 \cdot 1/k)[(1 + B).y + (exp(y) - 1)]\} \cdot dy. \right.$$

This may be written compactly as, $\int_t^{t_0} H\, dt = I = \left. \int_0^{\hat{y}} dy \middle/ \{1/\beta - f(y)\} \right.$ $\quad$ ---------------- (8)

Where, $\beta = H_0 \cdot 1/k$, and, $f(y) = (1 + B).y + (exp(y) - 1)$.

Now, the maximum value of y, $\hat{y}$ will be realised when $\eta = 1$.

Hence, if we set Δt_x to the accepted age of the universe, i.e. $0.43272*0^{\wedge}18$ s, then, $\hat{y} = 0.35313$.

All of equations (6) through (8) suffer the same fate when the magnitude of the upper limit of the integral is substituted in the denominator, for then it becomes zero, the value of the integrand at that point and hence the whole integral becoming infinite.

To avoid this impasse we proceed as follows:

Let, $L = {}^1\!/_\beta - f(y) = {}^1\!/_\beta + 1 - g(y)$, where $g(y) = (1 + B).y + exp(y)$.

The function is now expanded about y as a Taylor series:

$$L(y - h) = L(y) - h\, L'(y) + {h^2}\!/_{2!}\, L''(y) - {h^3}\!/_{3!}\, L'''(y) + etc.$$

For very small h, second and higher order terms in h may be considered negligible.

Therefore, to first order, $L(y - h) = L(y) - h\, L'(y)$.

Now, $L[(y - h) - h] = L(y - h) - hL'(y - h) = L(y) - hL'(y) - hL'(y) + h^2 L''(y)$.

Hence, $L[(y - h) - h] = L(y) - hL'(y) - hL'(y) = L(y) - 2hL'(y)$.

It is then easy to show that we now have a recurrence relationship of the form:

$$L(y - nh) = L(y) - nhL'(y) \text{, where } \mathbf{n} \text{ is an integer.}$$

Each element of this relationship represents the inverse of one of the ordinates associated with a numerical solution of equation (8). Integrating (8) by the Trapezoidal rule entails dividing the region of integration into a series of ordinates of fixed spacing and applying the formula for the Trapezoidal rule. In this procedure the bounding ordinates are divided by two. However, for sufficiently small **h** we may neglect this and, because of the form for $L(y - nh)$ posit that a good approximation to the magnitude of the integral in the vicinity of the singularity may be obtained by simply summing the ordinates and multiplying by the spacing, **h**, between successive ordinates, with the proviso that **h** is made sufficiently small.

The nth ordinate in this process is given by:

$$\frac{1}{[L(y) - nhL'(y)]} = \frac{1}{[L(y) + nh(1 + B + exp(y)]}$$

If the expansion is taken about, $\hat{y}$, then, for any **n** in the range $\mathbf{1 \leq n \leq N}$, the **nth** ordinate is given by $\frac{1}{nh(1 + B + exp(\hat{y}))}$, for, as noted previously, $L(\hat{y}) = \mathbf{0}$.

Further, each ordinate is multiplied by **h**, hence, a good approximation to the magnitude of the integral, **I** is given by:

$$I = \frac{\{1}{[(1 + B + exp(\hat{y}))]\}} \sum_{n=1}^{n=N} \frac{1}{n} \text{ --------- (9).}$$

There is no formula for evaluating the summation in this equation, for it may be summed to infinity. However, the function was expanded about $\hat{y}$ and the setting of N to progressively larger magnitudes moves the penultimate ordinate closer to the singular ordinate located at the' position' of $\hat{y}$; the magnitude of the integral is the natural logarithm of the frequency ratio described before.

For example, the following is calculated:

With **k = 0.8160679** and the time at which q attains the magnitude **of 0.286 (i.e. 13.715*10^9 yrs bp)**, it is calculated that $\hat{y} = \mathbf{0.353007384}$. For **N = 10^6, N = 10^7** and **N= 2*10^7** the integral has the values **11.5787, 12.9824** and **13.5214** respectively. Lack of sufficient computing power rendered the investigation of the effect of further increase in **N** impractical.

The numbers immediately above are the logarithms of the frequency ratios at the position of the penultimate ordinate in the set of ordinates in the numerical solution. It should be noted that in the vicinity of $\hat{y}$ the magnitudes of the ordinates change very rapidly, becoming infinite at $N = \infty$, or alternatively, when $L(y) = 0$. Simpson's Rule is not used here in the determination of the integral since it is considered to be incapable of capturing the behaviour of the function in the vicinity of $N = \infty$, unless a very large number of ordinates are determined; the large matrices associated with using this type of solution here are beyond the capacity of the computing facilities available to the author.

Now, it may be shown that, for photons, the blackbody radiation temperature, θ is given by:
$\theta = {}^{\theta_0}/_a$, where **a** is the scale factor. Strictly, this should be written: ${}^{\theta}/_{a_0} = {}^{\theta_0}/_a.$

It is customary to set a_0 to unity; but we retain the last form for we have shown previously
[6], that, ${}^{a_0}/_a = {}^{l_0}/_l = {}^{v}/_{v_0}.$ It then follows that, ${}^{v}/_{v_0} = {}^{\theta}/_{\theta_0}.$

Hence, ${}^{\theta}/_{v} = {}^{\theta_0}/_{v_0};$ in particular, $\left({}^{\theta_f}/_{v_f}\right)^4 = \left({}^{\theta_0}/_{v_0}\right)^4$, where, as noted before, the

subscript 'f' is reserved for conditions at the end of a free vacuon mode.

The temperature, θ_f of the heated baryonic matter at the conclusion of such a mode is given

by: $(\theta_f)^4 = \overline{F}\,(v_i - v_f).(v_f)^3$, and $\left({}^{\theta_f}/_{v_f}\right)^4 = \overline{F}\left({}^{v_i}/_{v_f} - 1\right) = \left({}^{\theta_0}/_{v_0}\right)^4.$

Now, neither of the two frequencies at the beginning and end of the jerk are known, but, from
the above equation we may determine their ratio, for $v_0 = 2.681\ T\ Hz$ and $\theta_o = 2.725\ K.$

Hence, ${}^{v_i}/_{v_f} = 1 + \dfrac{1}{\overline{F}} \left({}^{\theta_0}/_{v_0}\right)^4 = 1 + 2.2 * 10^{-5}.$

It is then concluded, in terms of cosmological time before the present time, that the size of the
frequency ratio calculated above is indicative that the time interval over which the jerk occurs
may be relatively very small. It is then reasonable to assume that the magnitude of Hubble's
parameter at the beginning of the jerk is little different from that at the end.

Whilst it is assumed that the trajectory of the deceleration parameter from the end of the jerk
until the present time is given by an exponential distribution, there are insufficient boundary
conditions to determine of the constant, **k**. The constant **B** may be obtained from the formula,
$q_0 = 1 + B$ provided that a reliable value of q_0 is available.

The value of q_0 quoted in [15] is, $-0.56 \pm 0.04,$ but this is an old paper dating from 2002.

The value, $-0.48^{+0.11}_{-0.13}$ presented by Vargas dos Santos et al [16] is much more recent, but is
quoted at the confidence level of **68%,** assumes flat space geometry and takes the initial
magnitude of the deceleration parameter to be **0.5**; this is inconsistent with the value of **0.286,**
determined earlier in this work.

Accordingly, given the trend of the values quoted for q_0 we assume a value of **-0.4**; it then
follows that, $B = -1.4.$

For the distribution assumed, the gradient of the deceleration parameter, dq/dt is given by

$-k.\exp(k(t_0 - t))$. The analysis cannot be continued unless a value of $\mathbf{k}$ is forthcoming. Hence, we assume that the gradient of the trajectory of the deceleration parameter at the termination of the cosmic jerk is the same as that at the beginning of the jerk. The justification for this is that both trajectories occur in the sterile mode and are also associated with the attenuation of thermal radiation.

Accordingly, we take $k_f = k_i = 0.81607 * (10^{-18})s^{-1}$ and hence the distribution of the deceleration parameter between the end of the jerk and the present time is then given by:
$$q = \exp(0.81607 * (t_0 - t)) - 1.4.$$

Since the trajectories of the distributions between the jerk and the present time are slowly-varying and do not possess any singularities, we use Simpson's Rule to calculate the magnitude of the associated integrals.

For the continuation of that trajectory emanating from the singularity it is calculated that the magnitude of the integral between now and the time of the jerk is **0.411596**. This must be subtracted from the magnitude of the integral between the surface of last scattering and the present time in order to determine the frequency ratio between the surface of last scattering and the jerk.

Using the values of the constants noted above it is calculated that the ratio of the frequency immediately after the jerk to that at the present time is **1.4743**; hence the frequency at the end of the jerk is **3.953 T Hz,** and the corresponding blackbody radiation temperature, **4.017 K**.

Now, we have shown that the ratio of the frequency before the jerk to that after the jerk is very close to unity; consequently, the vacuum frequency before the jerk is then taken to be **3.953 T Hz.**

Hence, the ratio of the frequency at the surface of last scattering to that at the beginning of the jerk,

$$\nu_s/\nu_i = 3.9 * 10^{18}/3.953 * 10^{12} = 0.9866 * 10^6.$$

Also, $\ln\,\nu_s/\nu_0 - \ln\,\nu_i/\nu_0 = 13.5214 - 0.4116 = 13.11 = \ln\,\nu_s/\nu_i$.

The result, $\ln\,\nu_s/\nu_i = \ln(0.9866 * 10^6) = 13.802$, is determined without reference to the integral solution whilst the ratio $\ln\,\nu_s/\nu_i = 13.11$ was calculated from the integral solution for $N = 2* 10\text{^}7$.   Increasing the number of steps increases the magnitude of the integral and is equivalent to decreasing the interval between the ordinates and hence approaching more closely the temporal position where the ordinate becomes infinite.

Indeed, we may determine the time elapsed after the maximum positive value of the deceleration parameter $(q = 0.286)$ at which the surface of last scattering occurs for $N = 2*10\char94 7$, as follows:

From the substitution, $y = k\eta\Delta t_x$ we may write : ${\widehat{y}}/{N} = k.\Delta t$, where Δt is that fraction of

the time interval, $t_0 - t_x$, and is the time interval mentioned above.

Hence, for $N = 2*10\char94 7$, $\Delta t = {\widehat{y}}/{Nk} = {0.35300738}/{(2 * 10^7 * 0.8160679 * 10^{-18})}$

$$\text{i.e. } \Delta t = \frac{0.2167 * 10\char94 11}{3.154 * 10\char94 7} = 687 \ years.$$

It should be noted that this result corresponds to that for $N = 2*10\char94 7$—at which stage there is still a disparity between the numerical solution of the integral for the logarithm of the frequency ratio and that determined from the logarithm of the frequency ratio of that frequency determined by the simple thermodynamic argument developed earlier, and that at the jerk; correspondence between the two results will be obtained by the systematic increasing of N in the numerical solution, thus reducing (perhaps substantially) the above time interval.

It was noted earlier that the frequency of the vacuum becomes infinite at $q = 0.286$. This is obviously incorrect for in [5] it was shown, for this model that the inception of the universe occurred at a vacuum frequency of approximately $0.515*10\char94 28$ Hz. We then speculate that, in the vicinity of $q = 0.286$, the universe entered a free vacuon mode of unspecified extent which bridged the gap between the deceleration phase following the inception of the universe and the Critical Trajectory. Further, by comparing the magnitudes of the frequency ranges over which the heating at the surface of last scattering and the cosmic jerk occurred, it is posited that this frequency drop and the corresponding time interval must have been extremely small. Whilst there is no way to determine the extent of the heating of the existing baryonic matter during this phase it can be demonstrated that only a very small frequency drop can initiate very large temperatures. Assuming the vacuum frequency in the vicinity of $q = 0.286$ to be of magnitude $10\char94 27$ Hz, then the product $\overline{F} \, v_f^3$ in equation (5) is very large $(4.876*10\char94 37)$ and that only a very small drop in the frequency of the vacuum (of the order of 0.1Hz) is sufficient to engender temperatures of the order of $10\char94 9$ K, which temperatures are characteristic of that of nuclear fusion, and thus facilitating the formation the light elements from some of the hydrogen. However, such an event could not be observed directly since it would be obscured by the surface of last scattering, and, in any event, in this model the radiation from the heated hydrogen/light element mixture would be attenuated at the position of the surface of last scattering by a factor in the region of $10\char94 9$, which is the ratio of the frequency at the 'crossover' point and that at the surface of last scattering. Thus, the photon energy density at the start of the surface of last scattering (i.e. before the heating event which will raise the temperature by **3000 K**) would be characteristic of an equivalent blackbody temperature of the order of **1 K.** Whilst the fusion event may then not be observed directly, its existence may be inferred by analysis of the primary baryonic matter.

Given that the frequency drop is expected to be very small, then it is not unreasonable to assume that the trajectory of **q** after the heating was virtually coincident with the Critical Trajectory; consequently, the natural logarithm óf the frequently ratio at **q** = **0.286** as given by the solution of equation (9) with **N** hugely in excess of **10^7** will be very close to $ln\,^{0.515\,*\,10^{\wedge}28}/_{3.953\,*\,10^{\wedge}12} = 35.94$, with the accompanying very small time interval calculated using the procedure shown above.

Due to the limited computing facilities available to the author the actual magnitude of the time interval cannot be determined, but if a calculation similar to that previously performed, with **N** very much larger than **10^7** is made, it is reasonable to expect that it will show **Δt** to be extremely small.

From the solutions of equation (9) for the different values of **N** and assuming a linear variation in the magnitude of the integral, a very rough estimate for **Δt** in this case is **40μs**. Hence the time span of the free vacuon mode at the 'crossover' in the vicinity of **q** = **0.286** may be taken to be of the order of **2*Δt** ≈ **10^-4 s.**

Finally, it must be emphasised that the overall path shown in Fig1 develops from the past to the future, and not vice versa.

The time interval of the cosmic jerk

We may calculate, approximately, the time over which the cosmic jerk occurs as follows:

The magnitude of the Hubble parameter at the end of the jerk is determined to be, **2.824*10^-18 s^-1.**

It was shown in [6] that **H** and $\boldsymbol{v}$ are related by the expression, $dv/dt = -v\,H$.

Rearranging this and writing it in finite difference form gives: $\delta t = {}^{-\delta v}/_{vH}$.

Now, $(\boldsymbol{\theta}_f)^4 = \overline{F}\,(\boldsymbol{v}_i - \boldsymbol{v}_f).(\boldsymbol{v}_f)^3$, therefore, $(\boldsymbol{v}_i - \boldsymbol{v}_f) = \dfrac{(\boldsymbol{\theta}_f)^4}{\overline{F}\,(\boldsymbol{v}_f)^3} \cong -\delta v$.

It is noted that, $\delta \boldsymbol{v} = (\boldsymbol{v}_f - \boldsymbol{v}_i)$.

Also, we have shown previously that $\boldsymbol{v}_i \approx \boldsymbol{v}_f$ and so we may use either of these in the determination of the approximate time interval of the jerk.

Hence,

$$\delta t \cong \dfrac{(\boldsymbol{\theta}_f)^4}{\overline{F}\,(\boldsymbol{v}_f)^3} \cdot {}^1/_{H\,v_f} = \left(\boldsymbol{\theta}_0/\boldsymbol{v}_0\right)^4 \cdot {}^1/_{H\overline{F}} = [{}^{2.2\,*\,10^{-5}}/_{2.824\,*\,10^{-18}}\ \text{s}].$$

This is equivalent to **247000 years**.

In terms of the time interval to the jerk from the present time (**5 billion years**) the above duration represents **0.005%** of that interval.

On this basis it is doubtful that the astronomical observations which confirmed the existence of the jerk had sufficient resolution to determine that the jerk was not a true jerk in the sense that it was not instantaneous, but occurred over the time interval calculated above.

The attenuation of the radiation emanating from the surface of last scattering

It has been shown that, $\theta/v = \theta_0/v_0$; this may be rewritten as: $\theta_s/v_s = \theta_i/v_i$.

Hence, $\theta_i = \theta_s . v_i/v_s = 3 * 10^3 . 3.953 * 10^{12} / 3.9 * 10^{18} = 0.003\ K.$

This substantiates the comment made earlier that, for this model, the blackbody radiation temperature at the temporal position of the jerk which emanates from the surface of last scattering is verging on insignificant in its contribution to the blackbody radiation temperature of **4.017 K** immediately after the jerk.

We proffer the speculation that the radiation emanating from the heated gas mixture at the surface of last scattering and, to a very small extent the radiation emitted from the light element formation phase, forms a uniform background for the CMB, and upon which is superimposed the lumpy detail attributable to the radiation emanating from heated matter at the end of the jerk. The justification for this is that the radiation from both the surface of last scattering and the light element formation phase emanate from a background, uniform throughout the universe, whilst that at the end of the jerk has occurred when structures on all scales were present throughout the universe. We exclude any radiative contribution from point-like objects for the ethos of this model is that only average conditions are considered, and to incorporate such radiation would require that it be averaged over approximately 10^92 enclosures, hence reducing its effect essentially to zero. In any event, the temperature of these objects would render it impossible for them to be affected to any significant extent by radiation from sources which are only a few degrees above absolute zero. However, it is known that clouds of cold dust and gas exist throughout the universe, and such entities may absorb and re-radiate the energy released during the free vacuon phase that constitutes the cosmic jerk. The above remarks contradict the conventional wisdom that the non-uniform features seen in the CMB are the result of quantum fluctuations in the early universe.

From all of the foregoing we may infer that had the cosmic jerk not supervened between now and the surface of last scattering, then the CMB would have been substantially more diffuse, possibly more uniform and hence exhibiting features different from that currently observed. Indeed, the sparse nature of the CMB under these conditions would have resulted in the hiss from an un-tuned radio at the present time perhaps being almost inaudible without substantial amplification.

Further, in this work the Surface of Last Scattering is a misnomer. In view of the earlier discussion it would be more appropriate to call it the Surface of the Second Heating, or, more aptly, the Cosmic Flash, the third heating occurring at the cosmic jerk. The light element formation phase would have no such title, being unobservable as a separate entity, and hence, in the current orthodoxy, unknown.

The scale factor at the surface of last scattering

If the calculation of the scale factor at the surface of last scattering is based on the current temperature of the CMB, θ_0 and that at the surface of last scattering θ_s, then the scale factor, a_s at the surface is given by; $a_s = {\theta_0}/{\theta_s} \rightarrow {2.725}/{3000} \rightarrow 9.083 * 10^{-4}$.

The redshift, $z_s = {1}/{a_s} - 1 \rightarrow {10^4}/{9.083} - 1 \rightarrow 1099.9$.

Now, the above results are predicated on there being no intervening effect between 'now' and the surface of last scattering.

It has been shown in this model that the cosmic jerk is associated with a heating of the baryonic matter in space, five billion years ago. Hence, the above results can only be described as <u>apparent</u> for they take no account of the effect at the jerk.

Now, it has been noted previously that the scale factor and wavelength are related by: ${a_x}/{a} = {l_x}/{l}$, but, during the cosmic jerk we cannot employ the expression, ${\theta}/{a_x} = {\theta_x}/{a}$,

for this is a free vacuon phase, and the expression applies only to photons.

Hence, in order to determine the scale factor at the surface of last scattering we must use the frequencies of the vacuum, as follows:

${a_s}/{a_0} = {v_{ji}}/{v_s} \, {v_{jf}}/{v_{ji}} \, {v_0}/{v_{jf}} = {v_0}/{v_s}$. Setting a_0 to unity the scale factor, a_s, at the

surface of last scattering is given by: $a_s = \dfrac{2.681*10^{12}}{3.9*10^{18}} = 0.687 * 10^{-6}$.

Elsewhere, [5] it has been shown, for a universe containing $10^{\wedge}80$ baryons, that the current size of the space occupied is that contained within a cube of side $76*10^{\wedge}9$ light years. Hence, at the time of the surface of last scattering the corresponding size of the universe is calculated to be in the region of $5.22*10^{\wedge}4$ light years.

Epilogue

It is considered that the extensive explanations deployed throughout this work render further discussion superfluous.

References

[Note: Due to a computer malfunction, work prefixed by 'H' is only available in hard copy form by contacting the author at willwinnmf@aol.com]

[1] Measurements of Omega and Lambda from 42 high-Redshift Supernovae.

    The Astronomical Journal 517(2):565-586.

[2] Observational evidence from Supernovae for an accelerating universe and a cosmological constant.

    The Astronomical Journal 116(3):1009-1038.

[3] A Universe from Nothing

    L M Krauss

    Simon &Schuster, 2012.

[4H] The Dark Energy coexisting with the vacuum of a universe modelled as a set of identical, frequency-quantised, simple quantum harmonic oscillators.

    W M Fidler September 2013.

[5H] On the evolution of the mass/energy of a universe from the vacuum, modelled as a set of identical, frequency-quantised, simple quantum harmonic oscillators.

Addendum: on the nature of Dark Matter---- a Neutronic Ring Theory.

    W M Fidler July 2014.

[6] The structure and expansion of a universe obeying Hubble's law and whose vacuum is modelled as a set of identical, bi-modal, frequency-quantised, simple quantum harmonic oscillators.

    W M Fidler February 2016.

[7] The Baryon Oscillation Spectroscopic Survey of SDSS-111.

    Copyright holders: 2013, the American Astronomical Society.

    Item ID 38259.

[8] Hyperphysics-phy-astr.gsu.edu/hbase/Astro/wmap.html.

[9H] Non-dimensional continuous and discontinuous representations of the electromagnetic spectrum and all such radiation spectra governed by Einstein's, $E = mc^2$ and the Planck-Einstein-Schrödinger equation, $E = hv$.

W M Fidler September 2012.

[10H] A chronological sequence of some of the cosmological parameters associated with the vacuum of a universe, derived from the modelling of the vacuum as a set of simple, frequency-quantised quantum harmonic oscillators.

W M Fidler January 2013.

[11] The Road to Reality

R Penrose

Johnathan Cape 2004.

[12] The Routledge Critical Dictionary of the New Cosmology.

P Coles.

Routledge Inc, New York.

[13] Lattice/modified cluster model of the atomic nucleus.

J Montgomery & R Jeffrey

Unclear2nuclear.   2008.

[14] Neutrons become cubes inside Neutron Stars.

F Llanes-Estrada & G M Navarro.

Arxiv.org/abs/1108.1859.

M I T Technology Review, August 2011.

[15] The Dynamical Parameters of the Universe.

M Roos & S M Harun-ar Rashid.

arXiv:astro-ph/0209611V2.

October 2002.

[16] Constraining the cosmic deceleration/acceleration parameter with Type 1a supernovae, BAO/CMB and H(z) data.

M Vargas dos Santos, RRR Reis & I Waga.

ArXiv reprint 1505.03814

Journal Ref: JCAP0(2016)66.

[17] The Cosmological Constant

T Davis & B Griffen.

Scholarpaedia.org/article/cosmological-constant

October 2011.

W M Fidler November 2017

Appendix

The following computer programs are written in BBC Basic and may be used to investigate scenarios with different values to those of the parameters used in the present work.

```
10 INPUT H0,X,Q,TJ

11 Y=1

20 Z=LN(Q+EXP(TJ*Y))*H0/X

30 IF ABS((Z-Y)/Y)>0.000001 THEN GOTO 40 ELSE GOTO 60

40 Y=Z

50 GOTO 20

60 K=Z

70 B=-EXP(0.1577*K)

80 PRINT K;" ";B" "

81 DEN= 1-((1+B)*X+(EXP(K*X/H0)-1)*H0/K)

82 PRINT DEN;" ";X;" ";10*X/(H0*3.154*1.372)-1

84 IF DEN >0 THEN GOTO 85 ELSE GOTO 87

85 X=X+0.000001

86 GOTO 11

87 TME=X/(2.177*3.154*(10^-2))

88 PRINT TME;" ";X" "

90 END

100 REM THE ITERATION OF EQUATION (2)
```

REM EXPRESSING H0 AS 2.177 WILL YIELD A VALUE FOR K WITHOUT THE MULTIPLIER 10^-18.

THIS MUST BE INCLUDED IN ANY SUBSEQUENT CALCULATION OF THE DECELERATION PARAMETER. FURTHER, ALL CALCULATIONS FOR K AND B,

TOGETHER WITH THE VALUE OF Q START WITH X=0.9. TJ IS THE TIME BP FOR THE START OF THE COSMIC JERK (= 5 BILLION YEARS = TJ = 0.1577,

CONSISTENT WITH DROPPING THE MULTIPLYING FACTOR 10 ^18).

THE PENULTIMATE LINE OF THE OUTPUT GIVES THE BEST VALUES OF K AND B, AND THE LINE ABOVE THAT SHOWS THE VALUE OF X CONSISTENT WITH

THE CHANGE IN SIGN OF THE DENOMINATOR.(A SPECIMEN SET OF INPUT VALUES IS 2.177,0.9,0.3,0.1577). THE LAST VALUE IN THE OUTPUT IS THE TIME BP

IN BILLIONS OF YEARS AT WHICH THE CHANGE IN SIGN OF THE DENOMINATOR OCCURS--AND HENCE RESULTS IN A NON-PHYSICAL VALUE FOR THE LOCAL

(IN THE TEMPORAL SENSE) HUBBLE PARAMETER. IT IS TAKEN THAT THIS INDICATES THAT THE EQUATIONS ARE NO LONGER APPLICABLE.

```
INPUT N,H0,DTX,BX,KX

DIM A(N)

J=0

FOR I=0 TO N STEP 1

 TRM = 1-H0*((1+BX)*DTX*I/N + (EXP(KX*DTX*I/N)-1)/KX)


 INV=H0*DTX/TRM


 A(J)= INV

 J=J+1

NEXT I


FL=A(0)+A(N)


X=FL

FO=0

FOR K=1 TO N-1 STEP 2

 FO=FO+A(K)

NEXT K


Y=FO

FE=0

FOR L=2 TO N-2 STEP 2

 FE=FE +A(L)

NEXT L


Z=FE

S=X+4*Y+2*Z


V=S/(3*N)
```

PRINT V

NUR=EXP(V)

PRINT NUR

REM THIS PROGRAM USES SIMPSON'S RULE TO EVALUATE THE INTEGRAL GIVEN BY THE FUNCTION DENOTED BY INV IN THE PRECEDING.

ANY OTHER FUNCTION OF THIS TYPE MAY BE EVALUATED BY CHANGING THE MAGNITUDES OF THE COEFFICIENTS IN TRM.

N MUST BE DECLARED AS AN EVEN INTEGER, FOR BBC BASIC WILL RETURN AN ARRAY WHICH HAS N+1 MEMBERS, AND HENCE AN EVEN NUMBER OF INTERVALS.

THE MAGNITUDE OF THE INTEGRAL IS GIVEN BY THE LAST NUMBER IN THE OUTPUT.

NUR IS THE FREQUENCY RATIO.

SPECIMEN INPUT: NUMBER OF STEPS,N; HO=2.18;TIME TO JERK FROM PRESENT TIME, 0.1577, BX=-1.4;KX=2.15. THE INTEGRAL MAY BE DETERMINED AT ANY OTHER TIME BP BY SETTING A

DIFFERENT VALUE FOR DTX, SOLVING, AND THEN SUBTRACTING THE MAGNITUDE OF THE INTEGRAL FROM THE PRESENT UP TO THE POSITION OF THE JERK. THE RESULT WILL YIELD THE RATIO: FREQUENCY AT THE PLACE OF INTEREST TO THE FREQUENCY AT THE JERK

.

```
10 INPUT KX,ZLIM,LT

20 TM =1-EXP(KX*0.1577)+EXP(ZLIM)

30 M =1/TM

31 SM=0

32 PRINT M

33 N=1


50 BT =1/N

51 PRINT BT

60 SM1 = SM+BT

61 SM=SM1

62 SM1= M*SM

63 IF N < LT GOTO 64 ELSE GOTO 90

64 N=N+1


80 GOTO 50

90 PRINT SM1
```

100 END

REM THIS PROGRAM IS REFERRED TO AS 'THE SUM OF THE INTEGRAL TERM PROGRAM' IN THE TEXT. KX IS THE CONSTANT IN THE EXPONENTIAL DISTRIBUTION

FOR THE DECELERATION PARAMETER, ZLIM IS THAT VALUE OF Z USING THE DELTA T AT WHICH THE MAGNITUDE OF Q BECOMES A MAXIMUM AND LT IS THE NUMBER OF STEPS

YOUR KNOWLEDGE HAS VALUE

- We will publish your bachelor's and master's thesis, essays and papers

- Your own eBook and book - sold worldwide in all relevant shops

- Earn money with each sale

Upload your text at www.GRIN.com and publish for free